AF583801

DEADLY SCIENCE

Numbers in nature

Contents

ADJUNCT ASSOCIATE PROFESSOR COREY TUTT OAM

DEADLYSCIENCE

DeadlyScience aims to provide Science, Technology, Engineering and Mathematics (STEM) resources to remote schools around Australia. So far, DeadlyScience has shipped more than 33,000 STEM books and resources to more than 800 schools across the country.

The organisation began when proud Kamilaroi man Corey Tutt found out that some schools in Australia were completely under-resourced and that Aboriginal and Torres Strait Islander children were discouraged from pursuing STEM because of this. DeadlyScience knows from personal experience that books and resources change lives and believes these kids deserve nothing but the best. Aboriginal and Torres Strait Islander peoples in Australia were the First Scientists of this land, and DeadlyScience is committed to preserving that history.

A world of numbers

Numbers are all around us. Most events and objects in the universe can be described mathematically because everything around us is made up of a number of atoms and molecules. From how our solar system works to how economies prosper, maths underpins everything. People have been using maths to study our world since the days of the very first scientists. But how? Read on to find out ...

DID YOU KNOW?

Mathematics is mostly about how patterns appear in the real world. Maths seeks to discover and explain patterns to help us understand life in our universe.

Number patterns

We can think about number patterns in many ways. Some patterns occur when we add or subtract numbers – these are called arithmetic patterns. Some occur when we multiply or divide numbers – these are called geometric patterns. Numerical equations either join things together and increase them (as in addition and multiplication) or split them apart and decrease them (as in subtraction and division). Equilibrium can be represented by the equal sign. In this way, we can see maths as the relationship between how we make less of something, more of something, or find a balance between things.

Predicting patterns

Patterns are only patterns if we can recognise, measure and predict them; otherwise, they may just be random events or coincidences. Science, Technology, Engineering and Maths (STEM) professionals measure patterns in categories such as time, distance, length, amount or cost. Some patterns are obvious, like odd or even numbers. Other patterns may be difficult to see and require lots of data collected over a long time, such as climate patterns like the occurrence of cyclones or hurricanes, or El Niño/La Niña climate cycles. Even things like waves, spirals and sequences are patterns that we can study and predict.

Symmetry

Symmetry is an important part of mathematics. Symmetry means an object or shape can be divided into two or more identical parts. Different types of symmetry explain how an object can look the same when divided. For example:

- A starfish has radial symmetry. Its body parts are arranged around a central axis, so it would look identical if divided through the centre into five parts.
- A hexagonal beehive has translational symmetry. It can be moved sideways without changing shape.
- The number 8 has mirror symmetry. It can divide into two parts that are mirror images of each other.
- A circle has scale symmetry. It does not change its shape when it is expanded or contracted.
- An equilateral triangle has rotational symmetry. It can be rotated around a fixed point and still look the same.

Masters of the Universe

What are two of the most important numbers in the universe? You might think of your favourite numbers, but in fact, two numbers that play very important roles are 1 and 0. Yes, zero is a number, even if it has no actual value! One and zero are important because they are used in binary coding to write the commands (or switches) that allow computers to run and perform complicated tasks. These numbers act as 'yes' or 'no' decision-makers in coding.

A slinky has helical symmetry. It can be twisted like a corkscrew and still stay the same shape.

DID YOU KNOW?

First Australians have a long history of seeking balance in ecosystems to conserve resources and ensure plants, animals and people survive. An example of this is totems. Totems play a major role in sustainability as individuals are given plant or animal totems they are not allowed to eat, which protects their totems from over-consumption.

FACT

The sand goanna is a totem of the Yuggera people of the Ipswich and Brisbane Valley regions of South-East Queensland.

FIBONACCI SEQUENCE

Spirals & sequences

FACT

Indian mathematician Acharya Pingala noticed the Fibonacci sequence in Sanskrit poetry in 200 BCE. It was named after Italian Leonardo Bonacci, who included it in his 1202 book *Liber Abaci.*

FERN FROND SPIRAL

Some patterns repeat themselves, some increase and some even decrease. If you look around, you will start to see numbers and patterns everywhere. You may even be able to use them to predict what will happen next.

Eye-spy spirals

A spiral is a curved pattern that starts from a point, moving farther away as it revolves around the centre. A spiral gets larger or smaller depending on whether you are moving to or from the starting point. Many spirals exist in nature, such as the fronds of a fern, the swirled shape of a nautilus shell, a spinning cyclone, and even our own spiral galaxy. You may hear people talk about spiral staircases or DNA as spirals. Technically, these are not spirals but helixes (or helices), which are three-dimensional (3D) curves.

DID YOU KNOW?

An optical illusion called the Fraser spiral illusion uses circles and shades of black and white to trick the brain into thinking it sees spirals.

Fantastic Fibonacci

Some numbers occur with great frequency in nature. The Fibonacci sequence is a pattern that appears often in both mathematics and in nature. The sequence occurs as each number is the sum of the two previous numbers. Starting from 0 and 1, the first few numbers in the sequence are: 0, 1, 1, 2, 3, 5, 8, 13, 21, 34, 55, 89, 144. What would be the next number in the sequence?

Trees, flowers, cyclones, shells, galaxies and even human faces can be linked to the Fibonacci sequence. Many seed heads, pine cones, fruits and vegetables display spiral patterns that follow a Fibonacci sequence when their parts are counted.

Waves of wonder

Waves are a type of pattern or sequence in nature, whether waves at the beach, or sound and light waves. All waves are just repeating patterns (or frequencies) of moving air, electricity and magnetism. There are two types of waves. In the first, the parts of the wave move back and forth, like sound waves – these are known as longitudinal waves. In the second, the parts of the wave move up and down, as in ocean waves – these are known as transverse waves. The light we see is an electromagnetic wave that is part of the electromagnetic spectrum. Electromagnetic waves can be small, like X-rays which fit millions of waves in a single centimetre, or large, such as radio waves that can be more than 100 m long!

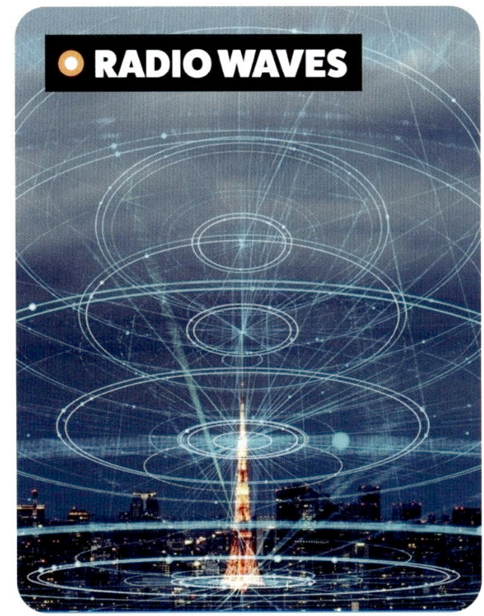

RADIO WAVES

FRASER SPIRAL ILLUSION

Tap & go

Did you know that some cultures have been using body tallying to count for thousands of years? It uses not only fingers, but also other body parts, such as hands, arms, and eyes. For example, counting starts on the little finger on the left hand (number 1), which the speaker touches or holds with the fingers of the right hand, moving past the ring finger (2), middle finger (3) and index finger (4), to reach the thumb (5). The counting then climbs the left arm, from the wrist (6), to lower arm (7), elbow (8), upper arm (9), and shoulder (10). Then, the speaker points to the neck, ear, nose, eyes, and crown of the head, before counting down the right arm from the shoulder.

Measurement

To know how much of something we have, we first need to count or measure it. This process of associating numbers with amounts of things, weights of objects, or repetition of events, is called metrology. This science of measurement quantifies objects or patterns in nature so we can make predictions about what will happen next.

Getting the measure of things

Counting and measuring things is just a way of comparing things you want to know with things you already know so that you can assign a value to them. Measurement quantifies the characteristics of an object or event, which can then be compared with other things or events.

MEASURING FOOD

Aiding with trading

Measurement compares an unknown quantity with a known, or standard, quantity and aims to ensure counts and weights are repeatable. Measurements are especially important in STEM fields, but they also help to ensure fair trade, help you bake a fluffy damper, or help you take the correct amount of medicine. Without proper measurements, it could be hard to tell whether you're buying 1 kg of flour or 1 cup of flour.

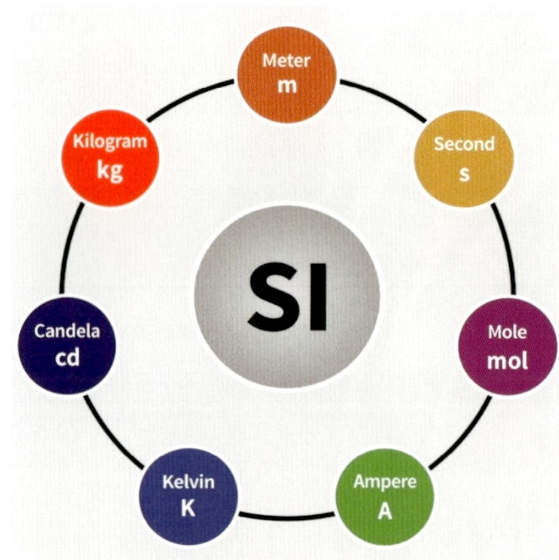

Maintaining standards

Comparisons of weights and amounts of things were once done by local agreements between trading partners, but there are now standards called the International System of Units (SI). Most people around the world use these standards in the metric system. When making trades, standard measurements make sure that the amount of something traded is the same for everyone.

DID YOU KNOW?

First Australian trade routes crisscross the country. The Yolgnu people of Arnhem Land, Northern Territory, even traded internationally with Macassan fishermen from Indonesia in the 1400s, long before Europeans arrived in Australia. Valuable items included pelts, ochre, wood, stones, mother of pearl, and weapons.

From senses to stopwatches

Astronomer Edwin Hubble said we explore the world around us with our five senses and call the adventure science.

We can take measurements using sight, sound, smell, taste, and touch; or we can use special devices or instruments to enhance our senses (such as microscopes or telescopes) or to record information (such as thermometers, tallies or stopwatches).

When measuring, we need to pay attention to accuracy, validity and reliability. To describe these concepts, we can use an example of measuring today's temperature.

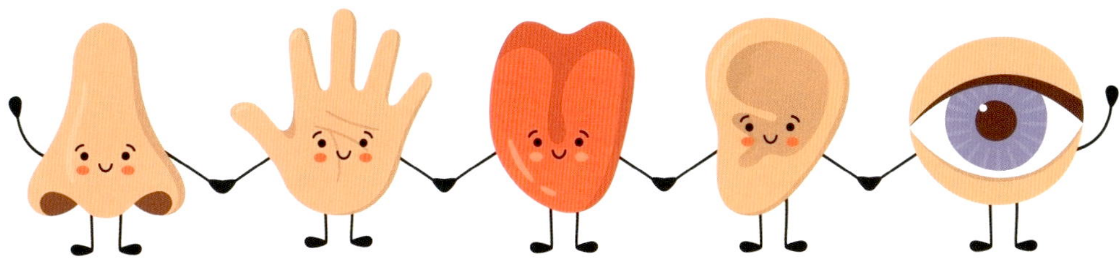

ACCURACY: An accurate measurement is the one closest to the actual value. For example, if today's temperature was 25 °C, the most accurate measurement would be the one closest to 25 °C.

VALIDITY: A valid measurement is one that is measuring the thing you're trying to measure. For example, measuring the temperature yesterday may provide an accurate measurement, but it is not a valid measurement of today's temperature.

RELIABILITY: A reliable measurement is one that is repeatable. If you measure again and get a totally different temperature, even though nothing else has changed, that is not a reliable measurement.

Tools of the trade

The device you use to measure depends on what you're measuring. It is hard to measure time with a ruler; it can be done, but it is not easy, efficient or accurate like using a clock or a stopwatch is. To measure temperature, you need a thermometer. And to figure out length, a ruler or tape measure is the most useful and appropriate tool.

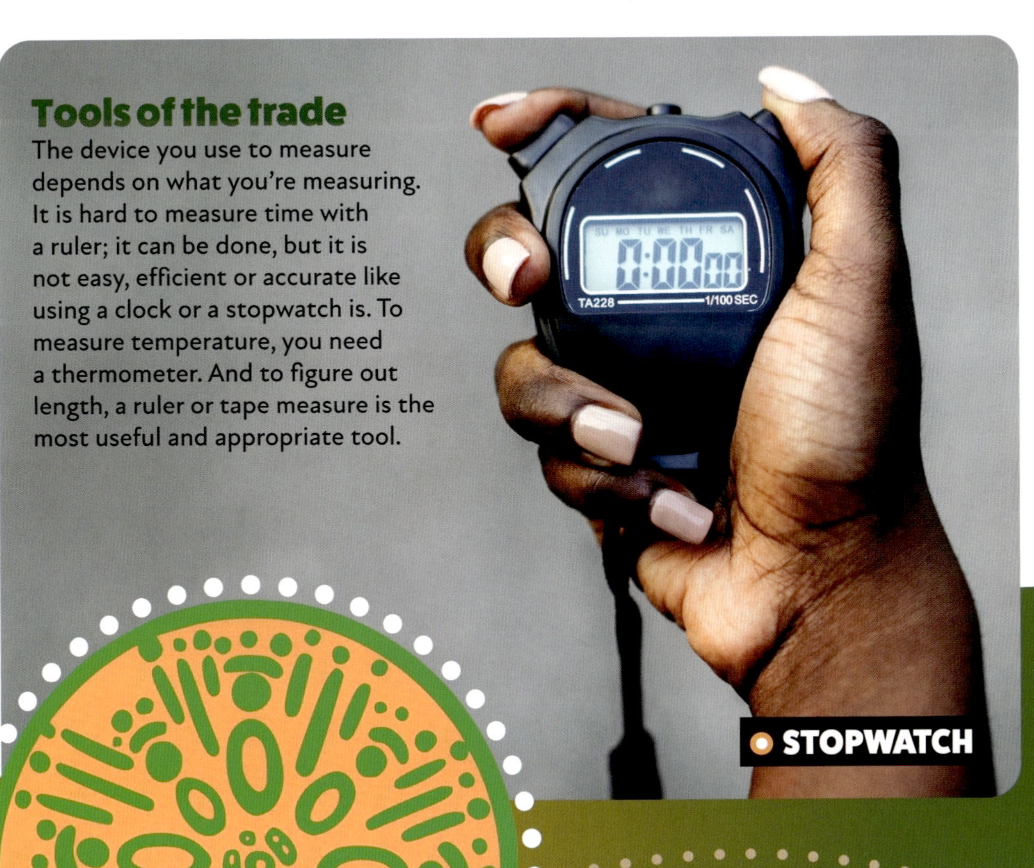

STOPWATCH

THERMOMETER

Tool time

Measurements are all around us, helping us learn about life and predict what could happen next. If you know the time, you can work out how long it is until lunchtime. If you know the temperature, you can choose what to wear so you're not too hot. If you know the Sun's position, you can work out which way is north.

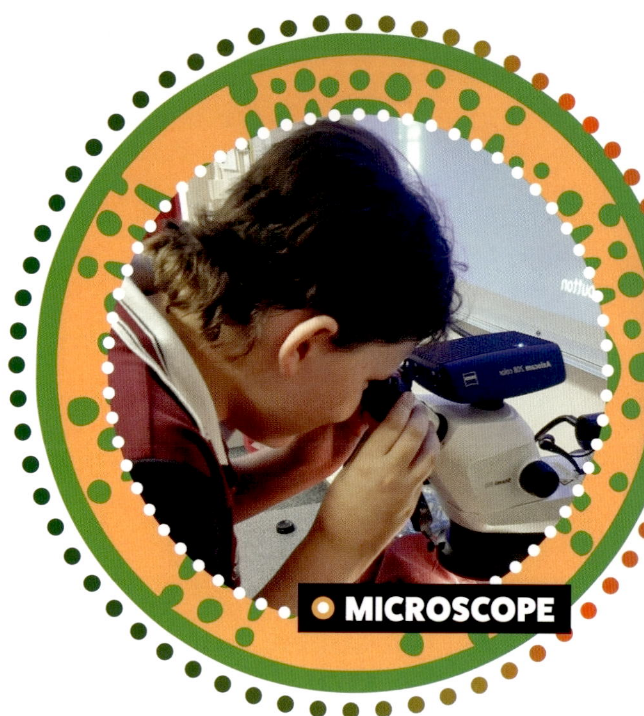

MICROSCOPE

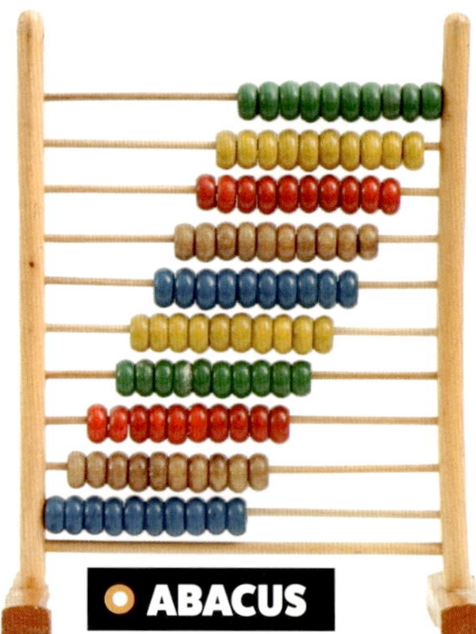

ABACUS

Look & learn

You may easily see a leaf on a tree, but to see the tiny cells the leaf is made of, you'll need a better tool than the human eye – a microscope. Tools we use every day help us improve our skills or our senses, and it is the same in mathematics. From simple concepts like addition or division using an abacus or calculator, to complex supercomputers and artificial intelligence that use equations and algebra to create algorithms, tools help mathematicians make sense of the information they gather.

Measuring devices can be as simple as using your fingers to count the number of objects you see, or as complicated as a telescope that measures the light intensity of stars. STEM professionals spend lots of time working with devices that make measurements. Some devices take a long time to learn how to use, such as magnetic resonance imaging (MRI) machines. Entire jobs are devoted to taking measurements, such as being a technician who makes sure machines work properly, clock-making or surveying.

SET SQUARES

SURVEYING

MRI MACHINE

All day, every day

We make measurements every day – some we notice, but others we may not. Some measurements are estimates, such as how much milk to add to your cereal, but others are precise and are taken with measurement devices, such as your alarm telling you it is time to wake up.

SOME EVERYDAY MEASUREMENTS

- Your speed in a car
- The time of day
- The distance you need to kick a ball to the goal
- The temperature inside or outside
- The amount of battery left on your phone
- The amount of sugar to add to a lemon myrtle cheesecake.

HUBBLE TELESCOPE

First Scientists measure up

Aboriginal and Torres Strait Islander people have been making measurements as scientists for many thousands of years. Measurements for information-gathering include noting the positions of stars and constellations to help with timing, travel, navigation and predicting seasonal changes. Observing temperature, wind and climate also helps determine when to take action, such as collecting food or making shelters, or when to expect fires or storms. Tracking plant and animal behaviours and movements enables people to figure out the location and numbers of certain species.

BARK HUT SHELTER

TRADING CANOES

DID YOU KNOW?

First Australians have long made measurements for experimenting, problem-solving and finding better solutions, such as strength-testing for glue to make weapons or implements, or testing the engineering of canoes to ensure that they float well.

Geometry

Geometry is one of the oldest forms of mathematics. The word comes from the Greek words 'ge' (which means earth) and 'metron' (which means measure), so it is literally 'to measure the Earth'. This branch of mathematics calculates space and distance, shape, size, perimeter and the position of figures and objects. It is used in art, architecture, engineering, motoring, machine building, robotics, and even space exploration!

DID YOU KNOW?

In many sports, players need to understand, estimate and account for speed, direction and angles to help them aim the ball or score. Luckily, geometry helps with all of these skills.

Exploring surfaces

Studying points, lines, planes, distances, angles, surfaces, curves and shapes is all part of geometry.

POINTS: The smallest most fundamental space, like a dot.

LINES: The shortest straight distance between two points.

PLANES: When three or more points are connected, the shape is called a plane.

DISTANCES: A measurement of length between points.

ANGLES: The inclination, or corners, between two lines that meet.

CURVES: Any non-straight connection between two or more points.

SHAPES: Objects formed by surfaces, lines and angles.

Lines and angles

A line is a straight object that has no real width. It can be any length, even infinitely long! A line can be drawn from left to right (horizontal) or top to bottom (vertical). Lines can be parallel (running alongside each other but never touching), perpendicular (crossing each other at 90°) or transversal (crossing two other lines at different points). Angles are the shape that is formed when two lines meet at a single point.

Angles are measured in degrees (°) or radians. A complete rotation is equal to an angle of 360 degrees.

ANGLES CAN BE:

- acute – when the inclination between the arms is less than a right angle, or less than 90 degrees.
- obtuse – when the inclination between the arms is more than a right angle.
- right – when the arms form an angle of 90 degrees.
- straight – when the arms form an angle of 180 degrees between them.

Angles, lines, and shapes form the structural basis of any building or bridge.

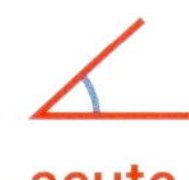
acute

right

obtuse

straight

Tracking through time

Aboriginal and Torres Strait Islander people used geometry to understand volumes of water. According to Dr Tyson Yunkaporta from Deakin University, geometry was also used to calculate time, based on the angles and position of the Sun, Moon and stars at different times, which helped First Australians predict changing seasons and weather. Other uses of geometry came in the form of reasoning and tracking skills. In oral traditions, formulas and proofs may not have been written down, but concepts of transformation and rotation were used to gather information about the world, such as the movements and directions of animal tracks over the landscape, or to make predictions about distance and time when following an animal or travelling.

Shapes in nature

Circles, hexagons, and triangles are very common shapes found in natural objects. Hexagons appear in beehives and some rock formations, while triangles can be found in flowers, birds' beaks, the ears of animals, and shells. Circular shapes are common in marine invertebrates, and eggs are oval-shaped ovoids. From tree trunks to tessellated rocks to snowflakes and hailstones, shapes in nature are everywhere.

Shapes in nature have also been borrowed and used by humans in engineering for hundreds of thousands of years, helping us create useful utensils, tools and more complicated structures such as homes and skyscrapers.

Shapes in art

Geometry and art are also very closely linked. For example, perspective in painting tries to make two-dimensional images look three-dimensional. The use of dots, points and lines is important to various art movements, such as pointillism and First Nations dot painting.

DID YOU KNOW?

As the Dark Emu in the sky changes over the year, First Australians use pattern recognition and geometry (reflection and rotation) to connect to an emu's life cycle. From this constellation, they can tell when it is a good time to hunt for emu eggs, leaving some in the nest to hatch.

EMU

Guriri means emu chick in the Marra language of the Gulf of Carpentaria, NT.

Trigonometry

The word trigonometry comes from Greek words for 'triangle' and 'measure'. It is the relationship between angles and lengths in triangles. Third-century astronomers' work in geometry led to trigonometry, which is used to solve complex problems. Today, it is used in surveying to calculate the lengths, areas, and angles between objects on land, in the oceans and in the sky. It is also used in music theory, architecture, electronics, medical imaging, seismology, meteorology, image compression, engineering, cartography, and game development.

ESCHER AT WORK

Tessellated trickery

Graphic designer M.C. Escher was well known for his mathematical-inspired artwork of repeated patterns and shapes. Using what he knew of geometry and perspective, Escher created artwork and images that seemed to defy reality. From waterfalls that shouldn't exist, to people walking down but also seemingly up stairs at the same time, Escher was a master at manipulating the rules of geometry and perspective to trick our eyes. Escher found particular beauty in strange, complicated tile patterns, called tessellations, of different shapes. His works included animals such as lizards, fish and birds. Tessellations also occur naturally in geological formations such as Tasmania's Tessellated Pavement.

TRY IT YOURSELF

Estimate how far away from you an object is.

Trig tricks

You can use trigonometry and the method below to estimate the distance of objects.

1. Close or cover your left eye and stretch out your arm. Put your thumb up and place it over the object you want to measure the distance to.
2. Now, switch eyes, so that the open one is now closed, and the closed one open. You'll notice your thumb has moved. Take a guess at how far your thumb has moved.
3. Now, multiply that estimate by ten – and that is your expected answer. This inexact method works because the distance from your eyes to your outstretched thumb is roughly ten times the distance between your eyes.
4. Another trick for estimating long distances is that if you know your own height, imagine yourself lying down, and count how many of you would fit in that distance.
5. Impress everyone with your new skills in estimating!

Astronomy uses trigonometry to locate and track celestial objects and to predict eclipses.

Sextants were used in navigation to measure the angle of the Sun or stars with respect to the horizon.

Tessellations also occur naturally in geological formations, such as in the Tessellated Pavement of Tasmania.

Golden Ratio

P O Q

a b

a + b

$$\frac{a}{b} = \frac{a+b}{a} = 1.6180339887... = \Phi$$

Ratios rule

A ratio is a type of relationship between two things that calculates how many times a part fits inside the whole. One significant and common pattern in both mathematics and in nature is known as the golden ratio.

If you draw a line and break it into two smaller, unequal parts (let's call them a and b as in the illustration at top left), the line's total length when divided by the larger part (a) will be the same as when you divide the length of part a by the smaller part (b) because these two lengths follow the golden ratio (a ratio of about 1.618).

This ratio is associated with our understanding of beauty, or aesthetics. This simple yet powerful ratio can be seen in artwork that has survived on rocks and caves for thousands of years and in more modern artwork, too. It is visible in the structure of shells, the human body, the shape of hurricanes and spiral galaxies, elephant tusks, starfish, sea urchins, ants and honeybees. For some reason, our brains find the golden ratio appealing, so we are drawn to this ratio. Amazing, right?

GOLDEN RATIO

The sweetest number

One of the most important numbers in maths is pi, represented by the symbol π. Pi is a very simple but complex number related to geometry – and it turns up all over the place! It comes from the ratio between the circumference and radius of a circle. (In fact, it is the circumference divided by the radius).

The circumference is the distance around the outside of a circle, and the radius is the length from the middle of the circle to the edge. This number is the same for any circle, however big or small.

Pi begins with 3.14159, but it goes on ... and on ... and on forever! We know the first part of π very well – some people can even recite this number up to thousands of places – but we will never know that last number, because it is infinite.

Yes, pi is pretty sweet (pardon the pun), but it is considered a special number in mathematics because it relates to cycles, like a circle that keeps going around. Whenever you see a pattern that repeats itself, chances are π will be involved. Based on what we know about patterns, geometry and trigonometry, you can understand why pi is such an important number.

Diggin' the data

When we see patterns in nature, we can begin counting those patterns. The way we count them is through data collection. What we do with that collected data is called data analysis.

Making sense of data

Data can tell us a story about what nature is doing. We can use numbers, maths and patterns to read the story nature is telling us and make sense of it and our world. Once data is collected, we can retrieve values, find data points, and establish a baseline (or normal state) to compare things with. A range of tools – including statistics, visualisations, formulas and algorithms, computations and categorisations – help us look at, analyse, and understand data.

Visualising data

Data representations help us view data in ways that relate one number (or variable) to another. Visual representations of data can take many forms, either concrete or more abstract. Data visualisation is any form that tells the story of the data.

Infographics are powerful ways to present large amounts of information in simple-to-understand ways. Computer animations, diagrams, equations, stories and even objects can also provide useful data visualisations.

Data about things like the positions of stars, historic climates or events can be contained in stories and artwork. Even the rings within a tree's trunk contain data that reveals the tree's age and the atmosphere it grew in. Your DNA contains data about your ancestry and how humans migrated over the landscape.

GRAPH
One of the easiest ways to collect and compare data is in the form of a graph or table. But data doesn't have to mean graphs or spreadsheets.

TORRES STRAIT ISLANDER DANCE

Statistically speaking

Statistics is an area of maths that looks at the collection, organisation, analysis and presentation of data. When you have large amounts of data, statistics help make sense of it. From simply finding an average value to more complicated tools and methods, statistics tell a story about what we are measuring.

We can use statistics to figure out what has happened or to compare and contrast sets of data. These number facts help us get a sense of what might happen in the future. For example, by comparing past data to recent data, we can see that temperatures are hotter than normal. If we have good long-term datasets, we should be able to predict whether tomorrow will be colder or hotter.

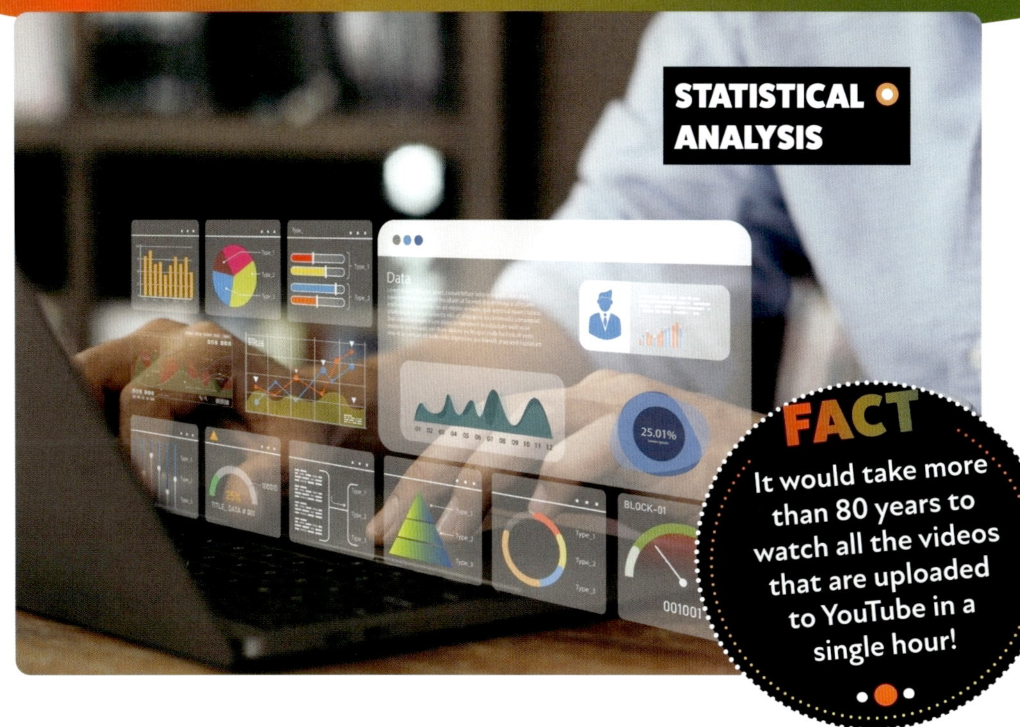

STATISTICAL ANALYSIS

FACT

It would take more than 80 years to watch all the videos that are uploaded to YouTube in a single hour!

YIDAKI

DID YOU KNOW?

Beat and rhythm are mathematical concepts, too. Music is a type of data made of repetitive sections called measures, each with an equal number of beats. Beats are mathematical divisions of time, so when you dance or play an instrument, you can recognise and count these sound patterns.

Big data

Have you heard of 'big data'? We live in the information age, and our internet-connected world produces huge amounts of data, most of which is collected and stored by the devices or services we use on a daily basis.

With ever-increasing technologies, mathematicians are gaining access to more data than ever before.

Increasingly sophisticated computing and algorithms, and new mathematical techniques, allow mathematicians and statisticians to collect, sort, understand and present information drawn from this database of big data.

Questioning the data

Big data collection must consider four questions.

1. How much data is there? In some cases, there may be more than we have the capacity to store electronically.
2. How varied is the data? Highly variable data gets harder to link or make sense of.
3. How fast is the data arriving?
4. Does all this data even need to be collected or should some of it remain private?

Problem solving

Many problems can be broken down and solved by mathematics. Numeric equations and algebra help scientists answer questions such as how many planets there are outside our solar system, what atoms are made of, or when is the best time to plant or harvest crops. Engineers use algebra to determine how much weight a suspension bridge can take. Biologists use algebraic models to help predict how fast populations of animals will grow. Even Google Maps uses mathematical equations to determine how you can best avoid traffic. You might not notice it, but equations, algebra and algorithms influence our daily lives.

A balancing act

The word algebra comes from an Arabic term that means 'the science of restoring and balancing' – and when we use algebra, that's exactly what we're doing.

Algebra allows scientists and mathematicians to solve problems by using huge amounts of data all at once, rather than having to deal with just individual bits of data.

Mathematicians use letters in algebraic equations instead of numbers. In this way, the letters can represent any numerical value. While using letters, you can still do all the usual things you do in mathematics – adding, subtracting, multiplying and dividing. The value of letters is that they can be either constant (meaning they do not change their value) or variable (changing their value depending on what is being measured).

DAVID UNAIPON

DID YOU KNOW?

Ngarrindjeri man David Unaipon was the first Indigenous person to be published as a scientific author. He used the principles of mathematics to solve engineering problems, such as his 1909 redesign of the mechanism that was the basis for modern sheep-shearing clippers.

Super solvers

Mathematics and science are two fields that often go together because they both use calculations to explain what we observe in our universe – and sometimes even things we can't see because they're happening on a tiny, or quantum, particle level. Super-powerful computers that operate using quantum mechanics can now perform complex equations and solve problems extremely quickly. And at the heart of all that computing is the language of the universe – mathematics.

FACT

Algorithms drive the apps we use to identify bird calls, socialise with friends or order takeaway food.

Amazing algorithms

An algorithm is an equation that acts as a programming tool (or set of instructions) to solve problems that reoccur in mathematics, computer programming and computer science. The problem might be in analysing, searching for or sorting data. For example, when you use Google to research for your homework, an algorithm behind the scenes takes your search query term as its input. It then runs the term through a set of encoded instructions for searching a database (like the internet) and returns search links for you that are known as output – so quickly that you barely realise a mathematical equation is involved!

Many natural processes have been used as the inspiration for algorithms, and these are known by the rather long name of 'nature-inspired optimisation algorithms' (or NIOAs for short). Studies of the swarming behaviour of bees, the parasitic behaviour of cuckoos and even the working of neural networks in our brains have been used as the basis for NIOAs in computer coding.

VIRTUAL REALITY

SWARMING BEES

Code red!

Binary computer code uses just two numbers – 0 and 1 – with mathematical equations as its commands to switch actions on or off. It might seem hard to believe, but when you are playing a video game written in binary code, all the graphics and gameplay are being controlled behind the scenes by code made up entirely of long sequences of different combinations of zeros and ones! Newer types of code, like Javascript, are more sophisticated and help create virtual realities, but video games are really just numerical data being used and interpreted in different ways.

About time

Time is a fundamental property of our universe. As far as we know, time only moves forward, and curiously, as it changes, it also stays the same. Basically, time is complicated! But human measurements of the passage of time have been carried out for millennia.

HOURGLASS

Keeping track of time

From the motion of the Sun to the motion of atoms, to the months in a year, and just about everything in between, humans are timekeepers who like to keep track of our changing days and lives. Natural events like day becoming night or the motion of shadows across the ground as the Sun crosses the sky help us do this. But humans also make tools, such as sundials, calendars or watches, that allow us to more precisely record the movement of celestial objects, the change of season, or the passing of minutes, hours, days or weeks.

LAKE MOORE, WA, ON BADIMIA COUNTRY

DID YOU KNOW?

Complex stone arrangements have been found across Australia. These may have been used for navigation and timing, or perhaps indicated events like summer and winter solstices.

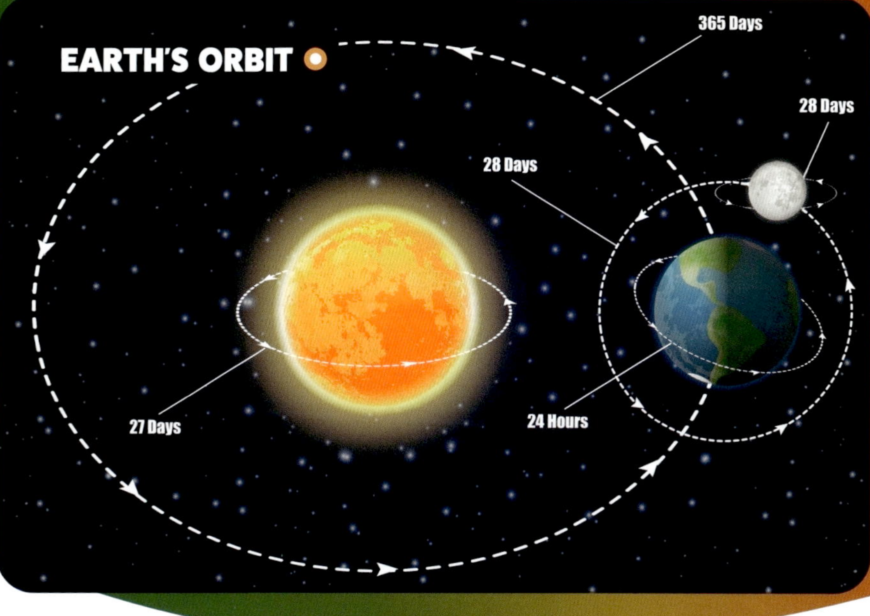

Telling the time

Some of the ways we measure time include the use of tools, such as:

- sundials, which follow the Sun's movement across the sky but only work during the day if the Sun is out.
- water clocks, which measure the time it takes to fill a known volume but the temperature of the water can change the volume (the same as for sand hourglasses).
- mechanical clocks, which use moving parts and cogs to count precise units like seconds. Issues include that clocks need regular maintenance.
- quartz clocks, which use the constant vibration of quartz crystals when close to electricity to measure very precisely but still need maintenance and batteries.
- atomic clocks, which use the precise vibrations of atoms but are complicated and very expensive.
- astronomical objects, like pulsars, which are used as timing devices in our universe. Pulsars aren't used to tell everyday time but to measure how precise other time measurements are.

Seeing it in the stars

You might not notice it, but the appearance and movement of the stars is an important factor in tracking time.

Our universe is very old, and our solar system and Sun within it. The stars you gaze at in the night sky are so incredibly far away that the light that reaches your eyes may already be thousands of years old. With a telescope, astronomers can see stars that are billions of years old! Constellations of stars, and the movement of them at times of the year, help people tell the time.

To track time, we compare it to something we know. When the Sun's position changes, we can tell that time has passed. We can also count a known number of time units – seconds, minutes, hours, days, months or years.

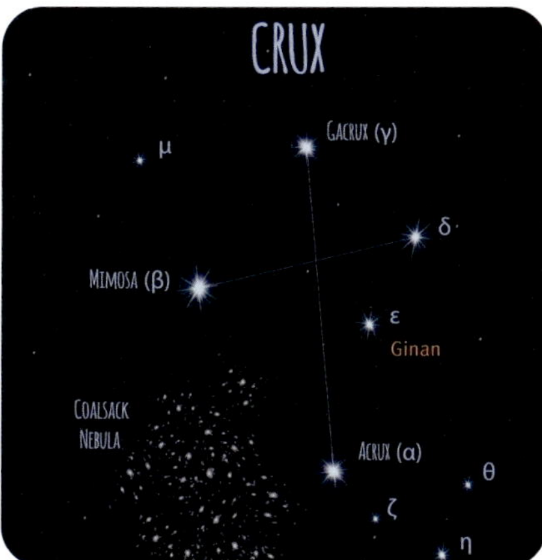

DID YOU KNOW?

Indigenous people were Australia's first astronomers who named the planets and objects they saw in the night sky. In 2018, a star formerly known as Epsilon Crucis in the constellation Crux (or the Southern Cross) was renamed to reflect the name used by the Wardaman people of the NT for thousands of years: Ginan. This star is some 228 light years away from Earth!

Time & change

In our daily lives, we use time to measure seasons, dates, time of day and smaller parts of the day (like when it is school time or dinner time). In STEM, we use time to measure how long an event takes. Scientists also do experiments to see how things change over time. We can work out how old fossils are, how old the universe is, and how many times a rainbow bee-eater flaps its wings.

RAINBOW BEE-EATER
This bird is known as birranga to the Noongar people.

LILLY PILLY FRUIT

Seasons, sun and star calendars

Calendars aren't always written down, but periodic measurements of timescales have been used globally for millennia. Seasonal calendars of First Australian cultural groups use the position of the stars and Sun in the sky or follow cycles of water, plant growth and animal reproduction as ways of knowing the seasons and when to perform cultural tasks.

Seasonal calendars vary depending on the region or Country where you live. The Top End may have six or seven distinct seasons, while other regions have two or three. It isn't just temperature that tells us the seasons. Changes in animal behaviour or the fruiting of plants do, too. Stories and knowledge held about the stars are part of First Australians' connection to the seasons and have been for thousands of years.

In addition to changes on Country, the moving stars or constellations, planets, Sun and Moon tell First Australians what to expect. Groups of stars or patterns that appear in the sky at certain times of the year are known as asterisms. Oral tradition recorded many asterisms, as well as 'indicator stars' that appear on the horizon and signify the change in season.

DHARAWAL SEASONS based on animal activity and the flowering of plants

WET — GORAY'MURRAI: Time of the eel. Kai'arrewan (coastal myall) blooms and fish are abundant.

HOT, DRY — GADALUNG MAROOL: Time of the kangaroo. Eating meat is forbidden, and weetjellan (lightwood) blooms.

WET — BANA'MURRAI'YUNG: Time of the quoll. Lilly pilly or satinash fruit ripens and falls.

COLD — TUGARAH TULI: Time of the burrugin (echidna). Eating shellfish is forbidden, and burringoa (forest red gum) flowers.

TUGARAH GUNYA'MARRA: Time of the lyrebird. East-facing shelters are built, and marrai-uo (Sally wattle or gossamer wattle) flowers.

MURRAI' YUNGGORAY: Time of the ngoonungi (flying-fox). Miwa gawaian (waratah) flowers.

JAN, FEB, MAR, APR, MAY, JUNE, JULY, AUG, SEPT, OCT, NOV, DEC

DID YOU KNOW?

For the Kaurna people of the Adelaide Plains, SA, an indicator star named parna marks the start of autumn (known as parnati) when the people move inland and make skin cloaks for warmth.

SPOTTED-TAILED QUOLL
The word quoll comes from the Guugu Yimithirr people of Qld.

Setting a date

Some of the earliest calendars marked the passage of days or seasons using sticks or stones, long before the complicated digital calendars we use today.

The important thing to realise about a calendar is that it is a cycle. Once we finish a year, the next year starts from the same date. Once a month finishes, the next one starts. The same goes for days and weeks.

Because time is cyclic, or periodic, scientists can make useful observations about what to expect or what might change. We know about seasons because changes in nature are observable at certain times in the year. We can also model, or predict, things like climate change by studying patterns that happened before and using computer algorithms and data analysis to research the probability of such things reoccurring or of other changes taking place.

FACT

Palaeontologists learn the timescale of fossils, petrified wood or ochre on rock surfaces by comparing them to the time it takes for carbon-14 to decay (known as a chemical's half-life).

TRILOBITE FOSSIL

CHECK YOUR SPEED

Distance over time

One example of a measurement of time is speed, which is simply distance (usually measured in metres per second, although in cars it is measured in kilometres per hour) divided by time. If you change your position and then measure the length between the two positions, you can also figure out your velocity. The greater the distance your position changes within a certain time, the faster you're going. Calculations of speed and velocity both use mathematical equations. To record the speed of boats and aircraft, a measurement of speed known as a 'knot' is used.

DID YOU KNOW?

Radiocarbon dating was developed in the 1940s to measure the decay of a radioactive isotope called carbon-14 in organic matter. It can date objects up to 60,000 years old, including First Nations rock art.

Navigation

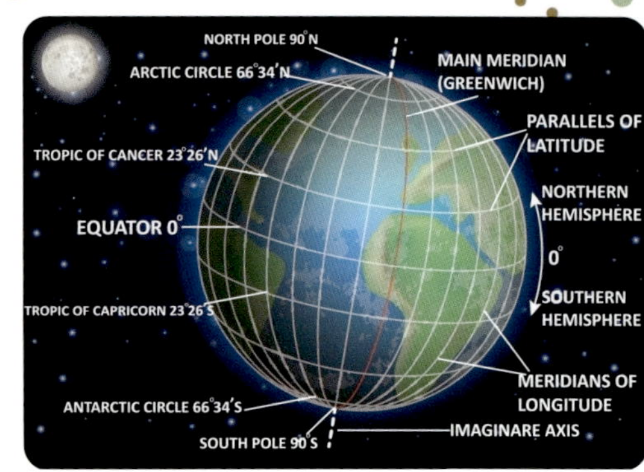

COMPASS

Wayfinding is a navigational process of moving from one place to another – over land or sea, through the sky, or in space. When setting off on a journey, to get where you're going, you'll need to follow landmarks or have a path or a map in mind to find your way. Celestial navigation, or following the stars, is a wayfinding method that has stood the test of time and helped First Australians and seafarers millennia before sextants, compasses or GPS devices were invented.

Latitude and longitude

Navigation can be seen as a comparison of your position with a known direction. When you set off on a voyage, you could choose to go left or right, or towards or away from something.

For land and sea navigation, these directions are called latitude (measuring north–south) and longitude (measuring east–west). When navigating in the air or in space, we must add height, or altitude. And in the ocean, we must add depth, which is measured in metres or sometimes in feet and fathoms.

TERMITE MOUNDS, NT

Termites can detect the Earth's magnetic field. They build their mounds with the thin side pointing north–south to keep them cool.

Finding a way

To make navigational measurements, your position is recorded in relation to a known point, such as the North Pole or South Pole. These points on the Earth can be easily pinpointed with a magnet, compass, or even, in Australia, a magnetic termite mound! In space and in astronomy, a similar concept called the 'celestial pole' is used. When measuring direction with a compass, it forms a 360-degree circle. North sits at 0/360°, east at 90°, south at 180° and west at 270°.

DID YOU KNOW?

In Litchfield NP, in the Northern Territory, you can see huge magnetic termite mounds that are useful for navigation as their thin edges always point north to south.

STORIES OF THE SKY

Follow those tracks!

Tracking is a specialised navigational skill that uses knowledge of local plant and animals, as well as complicated concepts in navigation, geometry and time, to find and follow something or someone. Tracking can include animal tracking, geographic tracking, and star tracking.

Skilled trackers use their observations and knowledge to orient and locate themselves or to find animals and water. First Nations artworks often include elements that represent the tracks and movements of animals or the location of water and food sources. In this way, they operate as maps that depict the songlines of Indigenous peoples.

Not within cooee

A cooee is a shrill call with a rising note at the end that First Australians use to communicate over great distances, as it carries far and echoes well. The sound was first recorded as coming from the Dharug language and was thought to mean 'come here'. Later, cooee was adopted into everyday use in the Australian bush, and was even used as a rallying cry to encourage soldiers to enlist in World War I.

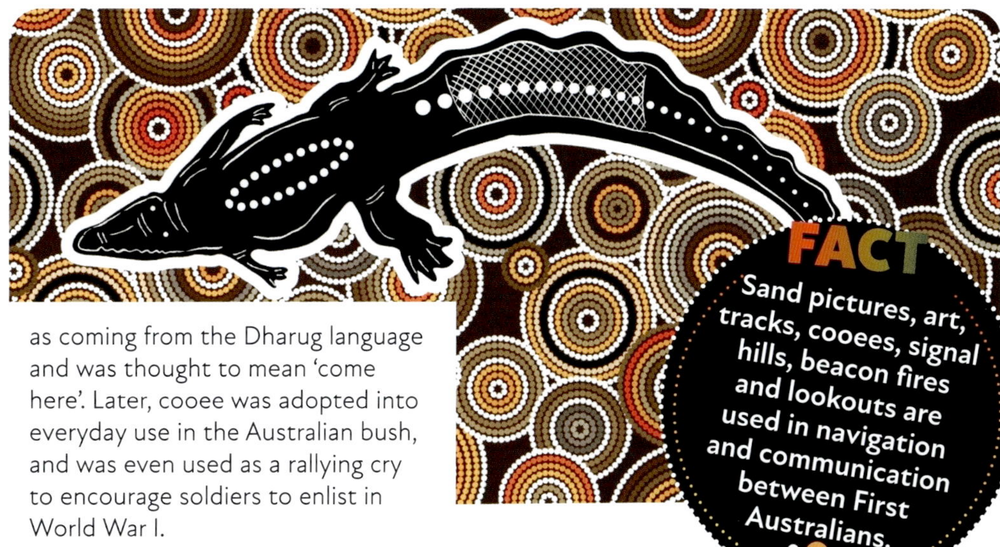

FACT

Sand pictures, art, tracks, cooees, signal hills, beacon fires and lookouts are used in navigation and communication between First Australians.

Triangulate a way

A trigonometry skill called triangulation can be used to find the source of a sound, like a cooee in the bush, if there are three listeners. Triangulation can be useful for locating a lost person calling for help, but biologists also use it to pinpoint the position of a calling frog or bird. It works by having three listeners surround the sound in a triangle, in roughly equilateral positions or points from each other.

Each person slowly approaches the noise, pointing to the direction where they think the sound is coming from. As the triangle gets tighter, the position of the caller becomes clearer.

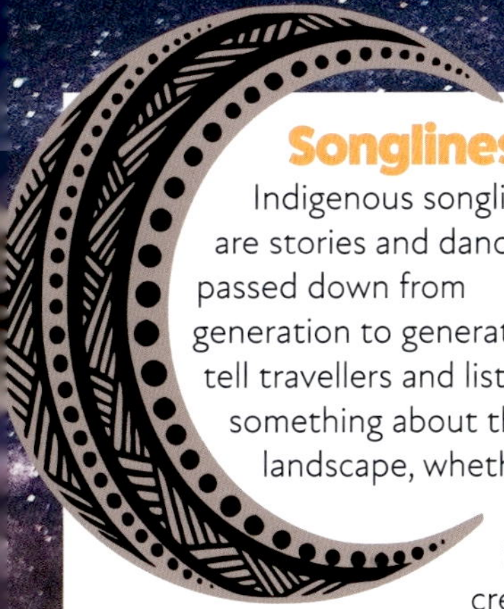

Songlines

Indigenous songlines are stories and dances passed down from generation to generation to tell travellers and listeners something about the landscape, whether about its creation or how to navigate it, or where landmarks and useful food and water are – or all of those things. These songlines stretch across the continent, and many of Australia's most prominent roads and trails still follow them.

DID YOU KNOW?

Songlines often include or follow sky maps that consider navigation by the stars. In this way, Indigenous knowledge about navigation is a cosmoscape that forms a vast story map of the land, the waters and the sky.

Tracking tips

DINGO TRACKS

Navigation and tracking use underlying number and mathematical skills, but they're also largely about paying attention – both to where you are and to where you are in relation to other things. Anyone can improve their navigational and tracking skills with a little bit of nature noticing, but a lesson from an expert also helps.

TURTLE TRACKS

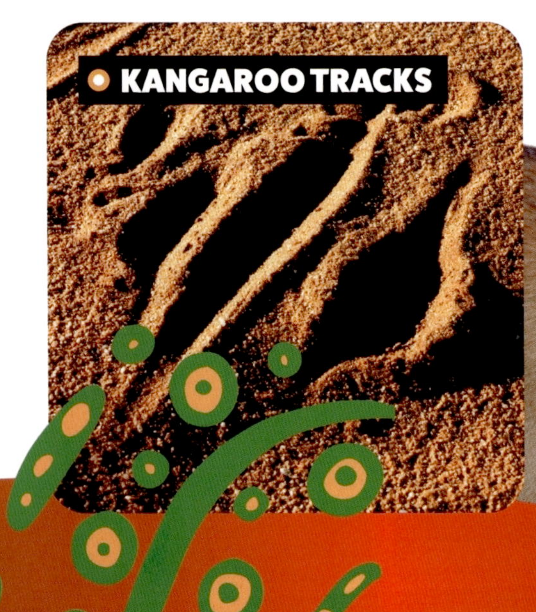

KANGAROO TRACKS

Animal tracking

The tracks, marks, traces and scats left by animals can help you find out what they are, where they have been, and where they're going. Most Australian mammals have distinctively shaped or coloured poo. Possum poo is a short, elongated pellet. Koala poo is greenish and smells like eucalypts, and wombat poo is grassy and shaped like a cube!

The movements animals make are encoded in their tracks, too, revealing how many legs they have, whether they have claws, and even whether an animal is injured or limping. A snake or crocodile slither looks very different to the hopping or scurrying marks of a kangaroo or a rodent, the track marks of a large bird like an emu, or the paw prints of a dingo.

A track record

In Walgett, New South Wales, you can find the Tracker Walford Walkway, which is named after famous Kamilaroi tracker Norm Walford.

Norm "Tracker" Walford was the last tracker officially employed in the state, and his main role was looking for people who were lost in the bush.

He worked around the state, using his classic mix of tracking skills. He even received a commendation from the Commissioner of Police for finding a man lost on Booka Booka Station, near Brewarrina, in 1957. The search for the missing man took three days and Tracker Walford came within 180 m of the body before it was found by a boundary rider.

DINGO *is the Dharug word for this species.*

BRUSHTAIL POSSUM *These possums are mudhay in the Yuwaalaraay language of north-western NSW.*

Star tracking

Knowing the position of stars and constellations can tell you where you are. One of the most famous celestial landmarks in the Southern Hemisphere is the constellation Crux, or the Southern Cross, which points south. At night, you can easily find your way south by a quick measurement of the Southern Cross and pointer stars. By day, you can find north by taking a measurement of the Sun (see page 30), but make sure you never look directly into it.

TRY IT YOURSELF

Knowing that the Sun rises in the east and sets in the west can help you find your way!

DID YOU KNOW?

In the Datiwuy language of northern Arnhem Land, NT, a game called waaiyn is played where a person makes an animal track or print in the sand and other players have to guess the animal from the track.

Geographic tracking

If you are ever lost in the bush, tracks, traces and symbols can help lead rescuers to your position.

Scouts, for instance, use trail markers to pass information on to others walking a trail. In the same way, First Australians have used symbols and markings in artwork and drawings in sand for thousands of years to communicate (there are more than 250 Indigenous languages including 800 dialects) or to pass on important information to others.

Sometimes, it might be passing on information about animals, explaining where fresh water can be found in the area, or warning of dangers in the landscape ahead.

Head for the hills

Changes in vegetation, areas where two types of landscape blend and merge (known as ecotones), or sites where rivers cross or change the way they flow may help you recognise which direction you're moving in. Features and landmarks like mountains help you judge where you are and more clearly see which way you might head. For example, if you walk down a hill towards another hill, you are likely to find a valley or a gully where water may be present.

MT LOFTY, SA

Map it out

The study of maps is called cartography. Maps help us navigate by referencing things such as rivers, roads, landmarks and boundaries, and recording the topology, or visible features, of a landscape. Maps are among the earliest and most important methods of sharing information.

First cartographers

Paintings, sculptures, body painting and message sticks can be used as maps, and sometimes include information about geography and topography. As we have seen earlier with navigation by the stars, First Australians often did not have to write or draw maps, because they were already in the sky; they just had to work out how to read them!

Skymaps

In many indigenous cultures around the world, the sky is seen as a reflection of the land and can be read as a map. This sky map includes information like distances, directions, way points and landmarks. For some First Australians, the Milky Way is seen as a river, and the stars can be used to show where waterholes or mountains are.

DOT ARTWORK

MESSAGE STICK

DIGITISED MAP

Maps usually include ...

1. A title telling you what area the map is covering.
2. A scale bar that explains the distance between elements.
3. A legend, explaining what the different dotted or wavy lines or colours mean.
4. A compass rose or directional arrows to show which way is north.
5. Grid references or coordinates.

Some maps also provide topographic information about landmarks or habitats.

TOPOGRAPHIC MAP

Types of maps

A globe is a three-dimensional representation of where we are in the world, while a two-dimensional paper map can help us locate where we are within a country, state or area. Digital maps, such as Google Maps, often also include terrain maps and street maps. They can tell us how to get from one place to another because GPS data figures out where we are and where we want to go. Other types of maps include songlines, which describe how to navigate through Country based on stories, landmarks and timing. Story maps help create a timeline of events that happen in books, films or history. And even mind maps help us navigate an idea in our minds and connect topics or ideas.

DID YOU KNOW?

Your skin can be considered a kind of medical map, because some of the symptoms you experience (such as spots or swelling) can help doctors know what might be going on inside your body.

Maths games

Just like maths, games follow certain rules to get to an outcome. And just like maths, they're really fun! We can use maths to help us understand games and even to win them, but we can also use maths to learn a lot about the decisions societies make and to make predictions based on games that we play.

WEAVING USES NUMBER PATTERNS

Game theory

Game theory came about in the late 1920s when mathematicians started using maths to make decisions about economics or social science. Mathematician John von Neumann decided to see what people, or in this case the players of a game, would do in certain circumstances. Probability was used to figure out what their decisions meant or to predict what decisions would be made. In maths games, we think about winners in terms of how they win and what they win. Some games are cooperative, but others can be played competitively by all players at once. In some games, no one wins. In others, some win more than others because the chance of winning is not random.

Snap it up!

A lot of card games – including Snap, Go Fish, Solitaire and others – are based on probability. In Snap, winners identify the highest number of similar cards and snap them! At the start of the game, there is a 1 in 13 chance of getting the same card – because each pack of 52 cards has four cards with the same number (i.e. there are four twos, four sixes, and four aces).

As the game progresses, the probability of pairs coming up changes because the number of cards left is smaller and some of the pairs may have already been removed. As the probability increases, so does the tension.

Games that use dice are also based on chance because we cannot control what the dice rolls.

CARDS GAMES USE MATHS

DID YOU KNOW?

Marn Grook is an Indigenous game that involves kicking and catching a stuffed ball made of possum fur and grass. Some people argue that Australian Rules Football was heavily influenced by Marn Grook.

AUSTRALIAN RULES

MARN GROOK FOOTBALL

FACT

Between 1942 and 1945, during WWII, a fighter aircraft made by the Commonwealth Aircraft Corporation was known as the CAC Boomerang.

Buroinjin

Buroinjin is a ball game played by the Kabi Kabi people of southern Queensland with a ball made of kangaroo skin (known as the buroinjin). To begin the game, the buroinjin is thrown in the air in the middle of the playing area. A player of one team is tasked with running as far as possible with the ball and crossing over a line at the other end of the field without being touched by an opponent. The easily understood rules allowed for social cohesion to form through organised play.

How dicey!

When using two dice, there is a much higher chance of rolling a seven than there is a two due to the larger number of possibilities of obtaining that number. For example, when you're rolling two dice, there is only one way to roll a two (1 + 1), but there are six ways to roll a seven depending on what number shows on the face of each die (1 + 6, 2 + 5, 3 + 4, 4 + 3, 5 + 2, 6 + 1). That's why the game of Yahtzee® is so challenging, because when you roll five dice at once, you have 7776 possible outcomes!

Number 7 Boomerang

Many First Australian mobs use boomerangs for hunting but also often use them in play; for example, trying to get a returning boomerang to land in a particular place, which requires a good understanding of the use of angles. In Australia's Tanami Desert, the Warlpiri people describe the number seven as wirlki, or a non-returning boomerang with uneven arms that does not form a right angle (like the one third from the top in the image above). It is sometimes called a number 7 boomerang. Boomerangs are highly aerodynamic, curved tools with a convex top and a thin, wide surface. Their streamlined design uses some of the same aerodynamic principles that were later used in the design of aircraft, but the mathematical formula for how they work was only devised in 1970.

Shadows of the sun:

Wamoon Public School

Students at Wamoon Public School, on Wiradjuri Country in central New South Wales, investigated how the Sun can help us tell time and directions by paying attention to the shadows of an upright object.

SUNSET OVER WIRADJURI COUNTRY

Materials

- A ruler, or a straight stick
- A few rocks or items that will not move when on the ground

Observations and predictions

Make some observations about where and how far the shadows will move over several hours. Can you guess where the shadow will be at a certain time? How would you test this?

WAMOON STUDENTS SET UP THEIR SUNDIAL

Method

STEP 1
Find a sunny spot with no shade. (Please be sun safe and do not look into the Sun!)

STEP 2
Place your metre ruler or straight stick poking up from the ground.

STEP 3
Place a rock where the shadow of the ruler ends.

STEP 4
Note the time for your reference.

STEP 5
Return every hour and place another rock where the shadow ends. Note the time for each observation you make.

Questions

Did the shadow move the same amount each hour?
What direction is the shadow pointing in the morning?

What this shows

As the Earth spins, the Sun appears to move across the sky. We can use that to help us tell the time. Since the Sun rises in the east and sets in the west, we can tell which direction the shadow will be based on what time it is.

COMMON WALLAROO

The Wiradjuri names for this species are walaruu and yulama.

DID YOU KNOW?

Animals that are most active at dawn and dusk are known as crepuscular species. If you're ever lost, following these species at sunrise or sundown will often lead you to water.

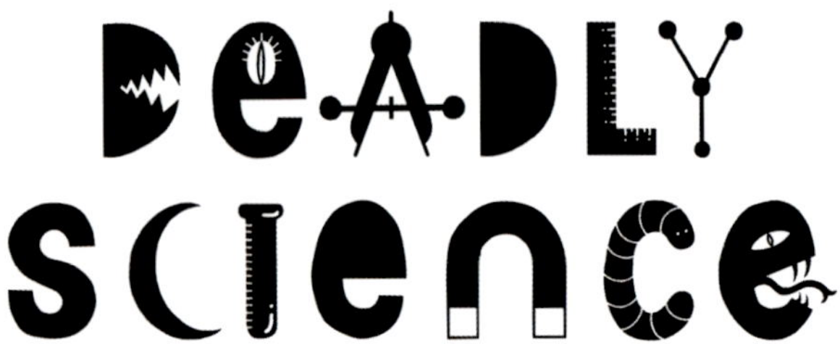

Numbers in nature

First published in 2024
© Australian Geographic Holdings Pty Ltd
52–54 Turner Street
Redfern NSW 2016

editorial@ausgeo.com.au
australiangeographic.com.au

Series Editor: Corey Tutt
Text: Jarrah Cain and Tom Gordon
Illustrations: Karina Jeffrey–Garawa Aboriginal Designs; Mim Cole, Mimmin Design

Commissioning Editor: Karin Cox
Layout: Karin Cox
Proofreader: Serene Conneeley
Cover Designer: Paul Hodge
Print Production: Andy Franks

AUSTRALIAN GEOGRAPHIC
Managing Director: David Haslingden
Editor: Karen McGhee
Head of Digital Content: Charlie Page
Licensing and Publishing Manager: Tom Bates

Printed in China by C&C Offset Printing Co. Ltd.

A catalogue record for this book is available from the National Library of Australia

Picture credits

Front Cover: Karina Jeffrey–Garawa Aboriginal Designs (all illustrations); Ethnografiska, Stockholm Sweden, WMC (Wikimedia Commons); polygraphus/SS (Shutterstock); Alec Trusler/SS; ChameleonsEye/SS; Florian Berger/US (Unsplash); wisely/SS; Kolonko/SS; YARphotographer/SS. **1:** DeadlyScience. **2:** Cherries/SS. **3:** Elena Abrazhevich/SS; Paggi Eleanor/SS; Florian Berger/US; Cast Of Thousands/SS; Ken Griffiths/SS. **4:** Africa Studio/SS; YARphotographer/SS. **5:** Karina Jeffrey–Garawa Aboriginal Designs; metamorworks/SS; Mim Cole. **6:** LookStudio/SS; Crystal Egan/SS; AydnPC/SS; Nana Margono/SS. **7:** Elena Pimukova/SS; Andrey_Popov/SS; Rawpixel.com/SS. **8:** DeadlyScience; PGL/US; Valerie V/US; Just dance/SS; Crissy Jarvis/US. **9:** Vadim Sadovski/SS; Michiru Maeda2003/DT (Dreamstime); Harden Sidney Melville/Public Domain. **10:** Robyn Mackenzie/DT; Zizou7/SS; showcake/SS. **11:** weter78/SS; Jason Benz Bennee/SS; Tony Baggett/SS; Shiraufas-art/SS. **12:** Pedro Ribeiro Simões/WMC; IgorZh/SS; sattahipbeach/SS; Framed/SS; FrankyLiu/SS. **13:** Dream01/SS; Africa Studio/SS; polygraphus/SS; Oksana Mizina/SS. **14:** ArtHead/SS; Ermess/DT; Pominoz/DT; BP.STUDIO/SS. **15:** Miriam FOTO/SS. **16:** Pixabay; Prachaya Roekdeethaweesab/SS; Gregory Johnston/SS. **17:** Chim/SS; PopTika/SS; Viesinsh/SS; Juila M Cameron/Pexels. **18:** zffoto/S; Karina Jeffrey–Garawa Aboriginal Designs; Jeremy Storey/Flickr. **19:** Adam Vilimek/SS; Siberian Art/SS; Jazziel/SS. **20:** AlecTrusler2015/SS; Treetstreet/CanvaPro; John Fung/SS. **21:** A_stockphoto/SS; Sergey Novikov/SS; Abrilla/SS. **22:** Triff/SS; Siberian Art/SS; ChameleonsEye/SS; Stanislav Fosenbauer/SS. **23:** Mykira/ DT; Karina Jeffrey–Garawa Aboriginal Designs. **24:** Louis Line/SS; Nynke van Holten/SS; Pawel Papis/SS; anne-tipodees/SS; EA Given/S. **25:** Lauren Cameo/SS. **26:** Ian Hitchcock/SS; Ethnografiska, Stockholm Sweden, WMC; Lucidwaters/DT. **27:** TunedIn by Westend61/SS; HammadKhn/SS; Lolostock/SS; Mariya II/SS. **28:** Lucidwaters/DT; Rawpixel.com/SS29. **29:** Nicholas Rjabow/SS; GustavMutzal/WMC; MzMacc/SS; Kolonko/SS. **30:** Cait Harding/SS. **31:** Wamoon Public School/DeadlyScience; Karina Jeffrey–Garawa Aboriginal Designs; Eric Isselee/SS. **Back Cover:** Mim Cole, Mimmin Design.

THIS PRODUCT IS LICENSED FROM AUSTRALIAN GEOGRAPHIC. AUSTRALIAN GEOGRAPHIC CONTRIBUTES 100% OF ITS PROFITS TO ITS REGISTERED CHARITY THE AUSTRALIAN GEOGRAPHIC SOCIETY (AGS) TO FUND ITS CONSERVATION AND OTHER GRANTS.
IF YOU WOULD LIKE TO KNOW MORE ABOUT AGS OR MAKE A TAX-DEDUCTIBLE DONATION, VISIT WWW.AUSTRALIANGEOGRAPHIC.COM.AU

AUSTRALIAN GEOGRAPHIC SOCIETY
Enquiries about sponsorship and donations:
02 9136 7206
Email: society@ausgeo.com.au
www.australiangeographic.com.au/society

AUSTRALIAN GEOGRAPHIC SUBSCRIPTIONS
Sales and customer enquiries: 1300 555 176
australiangeographic.com.au/product-category/subscriptions